DEAR READER...

Welcome to your
HORRIBLE SCIENCE ANNUAL 2013...
and what a slimy, creepy, star-studded
and chemically chaotic book it is.
Inside, you can visit a cave full of bat
poo, a lab full of lunacy and a dump
where cars get crumpled.
Plus you can learn how to squawk like
a bird, make a spud sprout
disgusting shoots, shine stars on your
ceiling and nip across the universe
in no time at all!
WARNING: it's magnetically charged,
unputdownable FUN!

CONTENTS

FREAKY FORCE

Finding this issue hard to put down? Maybe it's the invisible magnetic force coming from the pages! Discovering how to use this mysterious force has made the world the amazingly motorised place we live in today.

Electric motors, generators, body scanners, TVs and computer systems… they all use magnetism! But what IS it?

Mighty Magnetism FACT FILE

THE BASIC FACTS:

1. Magnetism is made by magnets. Each atom in a magnetic material is itself a tiny magnet.

2. Magnetism is the same force as the electric force made by electrons whirling around the atomic nucleus. The posh scientific name for the force is electromagnetism (e-leck-tro-magnetism).

3. What this means is that every atom is very slightly magnetic.

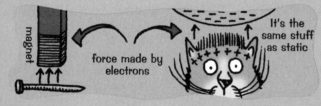

magnet

force made by electrons

It's the same stuff as static

THE SHOCKING DETAILS:

Question: But if atoms are magnetic and atoms are everywhere, how come everything isn't magnetic? How come you're not stuck to your bed in the morning? (No, you're not – it just feels like it!)

magnetic force

'mum'netic force

Answer: Remember, 'slightly' magnetic. You only notice a magnetic force if many billions of slightly magnetic atoms line up together.

The 'plane' facts

OK, so every atom is magnetic – and using metals with the most magnetic atoms of all has helped us make amazing machines. Talking of which, fancy making a plane? Only a tiny teensy one, but fun just the same!

DARE YOU DISCOVER...
how to make a magnetic plane?

You will need:
- a piece of tissue paper 2 x 1cm
- sticky tape
- scissors
- a thin steel pin
- a magnet (should be as strong as possible – or you could use several magnets in a line)
- a piece of thread 30cm long

All you do is:

1. Very carefully push the pin through the paper so it looks like a little plane (the paper being the wing).

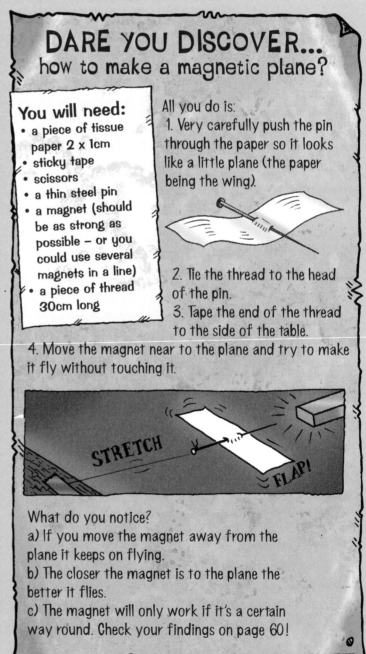

2. Tie the thread to the head of the pin.

3. Tape the end of the thread to the side of the table.

4. Move the magnet near to the plane and try to make it fly without touching it.

STRETCH

FLAP!

What do you notice?

a) If you move the magnet away from the plane it keeps on flying.

b) The closer the magnet is to the plane the better it flies.

c) The magnet will only work if it's a certain way round. Check your findings on page 60!

The inside story

How do you line up all those atoms? Surely you'd need a tiny pair of tweezers and loads of patience and it would still take for ever!

THERE'S GOT TO BE AN EASIER WAY...

Well, you'll be pleased to know that inside a magnet this lining up is done naturally by those nice helpful atoms.

1 Inside a magnet the atoms line up to form little boxes (about 0.1mm across) called domains (doe-mains). Inside these boxes the electrons can combine their forces to make what we call a magnetic force.

2 A magnet has two ends called north and south poles. Here their magnetic force is strongest. Two like magnetic poles (north–north or south–south) always repel – they push against each other. Two unlike poles (north–south) always attract – pull towards each other.

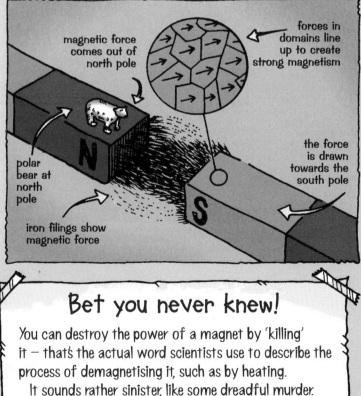

magnetic force comes out of north pole

forces in domains line up to create strong magnetism

polar bear at north pole

N

S

the force is drawn towards the south pole

iron filings show magnetic force

Bet you never knew!

You can destroy the power of a magnet by 'killing' it – that's the actual word scientists use to describe the process of demagnetising it, such as by heating.
It sounds rather sinister, like some dreadful murder. Well, if it was a crime, would you know how to solve it?

DARE YOU DISCOVER...
if magnetism works underwater?

You will need:
- a glass of water
- a magnet
- a paper clip

(No, you don't need a diving suit for this experiment!)
All you do is:
1. Plop the clip into the water.
2. Carefully place the magnet up against the outside of the glass.
3. Now try to use the magnet to bring the paper clip to the top of the glass without touching the paper clip and without getting the magnet wet.

GREAT CONCENTRATION IS REQUIRED

What do you notice?
a) It's easy.
b) You can't move the paper clip at all.
c) The paper clip only moves when you hold the magnet over the water – proving that magnetism works through water but not through glass.
Check what you find out on page 60.

A forceful tale

Magnets have a long history. According to legend, in about 2500 BC, a Chinese emperor used a rock containing magnetite (also called lodestone, or 'leading' stone) to guide his army through thick fog.

JUST FOLLOW ME, MEN!

OK – WHERE ARE YOU?

He probably used a sliver of the rock hung from a thread. Magnetite is naturally magnetic, so a sliver of it points north, which means can be used to make a compass.

CRAZY CAR CRUSHER

Magnets can lift cars! This marvellously magnetic recycling centre uses magnetism to move mangled motors.

1. When a car needs to be trashed, that doesn't have to be the end of it. Most of a car can be recycled – mainly the metal.

2. Mechanics drain away any oil and petrol and remove the battery and tyres.

3. Ever picked up paperclips with a magnet? Well, these 'electromagnets' can lift up a car! They are metal cylinders with coils of wire looped inside them (3a). When an electric current is passed through the wire, they suddenly become powerful magnets. Their strength depends on the size of the wire, the number of coils and the current's voltage. The electricity can be turned on and off with the flick of a switch, so the electromagnets can pick up huge weights and then – FLICK! – drop them.

4. Next, the car is shredded in a monster machine into little bits (4a).

5. A horseshoe magnet in the drum at the end of the conveyor belt picks up any 'ferrous' (iron-based) metals, such as iron and steel (5a), while the remaining 'fluff' (5b) falls into a skip.

6. Recycling centres also take kitchen appliances. Before theyre shredded, fridges and freezers have the poisonous, ozone-harmful coolant drained from them (6a).

7. Domestic rubbish goes on to a conveyor belt (7a). It passes under another magnet (7b) which grabs metal food and drinks cans and other ferrous items.

CLATTER

WHIRRRRR

CLATTER

MIGHTY MOTORS

Clean, powerful and fairly quiet... electric motors power all sorts of things, from the washing machine that cleans the grime off your football strip to the power drill used to fix the bed frame you broke when you jumped up and down on it!

The only time anyone notices electric motors is when they don't work or give their owners a nasty shock. But did you know that electric motors depend on magnetism and electricity working together? The force from an electric current can make a magnet move and this is exactly the principle behind an electric motor. Wanna know more?

MIGHTY MOTOR FACT FILE

NAME: Electric motor

THE BASIC FACTS:

1. Every type of electric motor uses electromagnetic force to make a wire loop move. Here's how...

wire loop turns

LOOPY INVENTION

battery

magnet

2. The electromagnetic forces in the wire and the magnets keep pushing against each other and this pushes the wire loop round.

3. The moving loop can be used to power the moving parts of a machine and keep it 'ticking over' smoothly.

THE SHOCKING DETAILS:
You'll find electric motors in loads of things... like an electric saw for cutting off the tops of dead people's heads so scientists can study their brains!

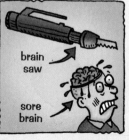

brain saw

sore brain

Dare you discover...
how magnetic forces work?

You will need:
• Two magnets

All you do is:
Put the magnets close together.

What happens...?
a) The magnets spring apart or pull together, depending on which way round they are.
b) The magnets are always drawn together.
c) The magnets can be placed together and you don't feel any force between them.

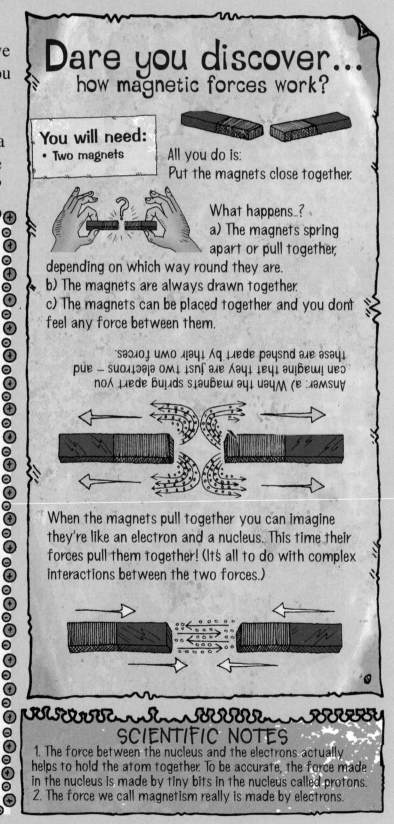

Answer: a) When the magnets spring apart you can imagine that they are just two electrons — and these are pushed apart by their own forces.

When the magnets pull together you can imagine they're like an electron and a nucleus. This time their forces pull them together! (It's all to do with complex interactions between the two forces.)

SCIENTIFIC NOTES
1. The force between the nucleus and the electrons actually helps to hold the atom together. To be accurate, the force made in the nucleus is made by tiny bits in the nucleus called protons.
2. The force we call magnetism really is made by electrons.

Dead brainy

In the early 19th century, the race was on to combine electricity and magnetism to make a working electric motor. The basic idea was thought up in 1821 by Michael Faraday. He built a machine to show this – and so created the first ever electric motor. In a scoop for Horrible Science, we've actually persuaded the great scientist himself to explain how it works. (This is quite amazing since he's been dead for well over 100 years!)

I REALISED THAT A WIRE CARRYING AN ELECTRIC CURRENT WOULD SWING AROUND A MAGNET...

electric wire

power supply   fixed magnet

...AND NOT ONLY THAT BUT A WIRE WITH AN ELECTRIC CURRENT COULD MAKE A MAGNET MOVE ROUND IT

power supply   fixed electric wire

magnet

IN EACH CASE THE MOVEMENT IS DUE TO THE ELECTROMAGNETIC FORCES FROM THE WIRE AND MAGNET PUSHING AGAINST EACH OTHER. OF COURSE, IF YOU TOUCH THE WIRE YOU'LL GET A NASTY SHOCK...

AAARGH! I NEARLY KILLED MYSELF!

M.F. HAS FORGOTTEN THAT HE'S BEEN DEAD FOR OVER 100 YEARS

He called this his 'homopolar motor'. We just call it 'brilliant'. But mad-for-magnetism Michael didn't stop there. He went on to prove that the electricity made by magnets, batteries and static was all basically the same thing. Electricity had really got motoring.

Spot the electric motor

Which of these familiar household objects contains an electric motor?

1. HUM!

2. TURN!

3. SPIN!

4. BLOW!

You guessed it – all of them!

1 Ever wondered why fridges and freezers hum? Chemicals are pumped around a pipe at the back (causing a hum), which creates a cooling effect. This pipe then passes into the main storage bit.

ZZZZZ
PUMP!   PUMP!
pipe full of cooling chemicals   back of fridge
power   pump

2 In a microwave oven, food is rotated on a turntable driven by an electric motor. The motor also drives a fan which helps to reflect microwaves onto the food.

yummy pizza   electric motor   turntable

3 The CD player uses a laser beam to scan tiny pits on the underside of a disc as it's spun around – by an electric motor.

4 Inside a hairdryer is a coil of wire that gets heated up by an electric current. A fan, driven by an electric motor, blows air over the coil. Voilà – hot air!

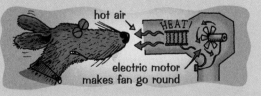

hot air   HEAT!
electric motor makes fan go round

CHOCO DYNAMO

Welcome Dr Chocolato's amazing boutique of sickly treats! How does this cruel chocolate maker use magnetism to power it all? Well, the idea is awful – but the science is sweet!

THE SHOP

1. Dr Chocolato's handing out free samples as the kids flock to his shop. He's clearly making a bundle, with cash flowing from the till (**1a**)!

2. The Doc's faithful assistant ape Cocoa the bonobo likes to take a taste of the famous chocolate people! But there's something a bit evil about Cocoa – he and Dr Chocolato hide a sinister secret behind the wall...

THE SINISTER SECRET

3. Unknown to all the keen customers, the whole shop is powered by slave children, pedalling to drive a dynamo on the false promise of a Christmas pudding! Their stationary 'penny farthing' bicycles have drive belts (**3a**) instead of chains, which, through a series of gears (**3b**), drive the giant dynamo in a secret room.

THE DYNAMO

4. The dynamo features a large coil of wires (**4a**) on a turning shaft (**4b**). They are housed in a giant magnet (**4c**). When the turning shaft goes round, powered by the cyclists, the wires in the coil experience a changing magnetic field from the magnet, so electricity is generated in them. This is called 'electromagnetic induction'.

5. Here, the orange arrow shows which way the shaft must turn (**5a**) and white arrows show the direction of electrical current in the coil and dynamo (**5b**). The magnetic field flows from the magnet's north to south poles.

6. The electricity is used to power the lights, the chocolate machine and everything else in the shop.

DROOL

Can you find the six choc bars sneakily hidden here?

LIGHTS & FRIGHTS

Magnets are great for sticking notes on the fridge, but they've got other uses, too, from creating theme park rides and woeful wartime weapons – but the biggest magnet is up in the sky!

Criss-cross Crazy

The Sun is magnetic. It is made up of loads of charged particles all rushing around. This movement creates an electromagnet, in the same way that an iron bar becomes magnetic when wire twisted around it has electricity passed through it (as you saw on page 9).

Sun's flares

son's flares

The surface of the Sun is criss-crossed by magnetic lines. Sometimes these lines pull matter out of the Sun and high into space in spectacular flares – solar flares – which are seen best during a complete solar eclipse. Solar flares send electromagnetic radiation hurtling through space towards Earth which can disrupt our radio communication, power supplies and electronic equipment. There was a massive solar storm like this in March 2012.

The coloured lights that appear in the skies over Earth's magnetic poles – called the Aurora Borealis and Aurora Australis – are caused by the Sun's solar wind hitting particles trapped by the Earth's magnetic field.

IS THAT THE SOLAR WIND?

NO, JUST TOO MUCH CABBAGE FOR TEA

A Big Lift

Maglev (*magnetic levitation*) trains have been developed in Japan and Germany. These trains are lifted off their guide rails by magnets, so they hover. No friction to slow them down means they can go super fast!

Here's how they work. As the train glides forward, powerful onboard magnets make electrons move in a reaction rail underneath the train. This creates an electric current which gives off a magnetic force that pulls the train forward.

Maglev technology has also been used to make lifts and some really cool theme park 'white knuckle' rides. Why not try persuading your parents to let you go on one for your science homework?

WELL, REMEMBER YOU INSISTED ON CHECKING MY HOMEWORK, DAD!

True or False?

1. The Canadian nickel coin is magnetic.
2. Magnets are used to pull metal splinters from patients' eyes after car accidents.
3. In Siberia, people fish with magnets.

Answers on page 60.

TERROR TALE

At the start of World War II, German naval mines kept blowing up British and Allied ships. Some of these mines used magnetic fuses, which were triggered by the steel hull of passing ships or submarines. So the British Admiralty looked into ways of avoiding these mines. One method was to try and make their ships undetectable by passing an electric current through coils wrapped around the hull – creating a huge electromagnet! But, according to eyewitnesses, the results of these experiments were far from what the Navy had expected. It is said that the test ship,

the Eldridge, and her crew became invisible! When they reappeared some of the crew were on fire, others had gone mad, some were missing, never to return, and a few had become stuck in the walls or floors of the ship!

HEY, WAIT GUYS, I'M STUCK

AARGH!!

AARGH!!

PUZZLES

THE CASE OF THE MURDERED MAGNET...
CASE FILES OF OFFICER LODESTONE, NYPD

Following a tip-off from some kids, we busted the flat of a top science teacher. Judging by the still warm cup of coffee, he had only been gone a few minutes. The flat was a mess and I felt dirty just being there.

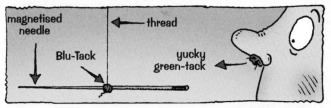

THE VICTIM

The magnet was face down on the table. There were no marks of violence on the body, but a quick check showed that the magnet had been killed — it had no magnetism whatsoever. Taking care not to smudge any fingerprints, I turned it over. The metal was ice-cold to the touch. Here are the possible murder weapons...

THE SUSPECT

GENERATOR

CANDLE

HAMMER

WARM CUSTARD

Your mission... is to find out how the magnet was killed. (Turn to page 11 to refresh your brain cells about killing magnets.) Was it by...

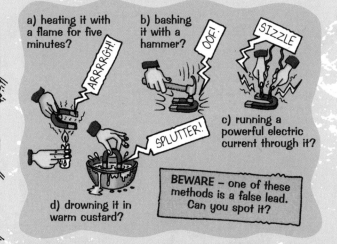

a) heating it with a flame for five minutes?

b) bashing it with a hammer?

c) running a powerful electric current through it?

d) drowning it in warm custard?

BEWARE — one of these methods is a false lead. Can you spot it?

Three of these methods would rearrange the atoms in the domains so the magnetic forces no longer pointed in a single direction. This would mean that the magnet lost its power. Read the case notes again... you may find a small clue.

Magic Needle

A nutty science teacher (what other kind is there?!) is showing some kids a barmy experiment using a sewing needle, a magnet, some thread, Blu-Tack, sticky tape, kitchen foil and a 1.5 volt battery. Read carefully what he does and try to guess what happens. (Do NOT cheat and try the experiment yourself as the battery can get a bit hot and scary!)

1. He strokes the needle with the magnet 30 times.
2. Then he fixes the needle to the end of the thread with a small blob of Blu-Tack in the middle so it hangs sideways in the air.

magnetised needle · thread · Blu-Tack · yucky green-tack

3. He sticks the other end of the thread to a table top with more Blu-Tack.

ANOTHER NUTTY EXPERIMENT! · Blu-Tack · table top

4. Now he takes the kitchen foil and cuts a strip from it. He folds this in half lengthways and then folds it again lengthways.
5. He uses sticky tape to fix one end of the foil to the positive (+) end of the battery and the other end to the negative (−) end. This makes a circuit for an electric current to run through.
6. Holding the battery horizontally, he puts the foil loop round the needle and moves the foil up and down without touching it.

What do you think happens?
a) The sewing needle starts to glow with a strange blue light.
b) The needle twists round spookily.
c) The needle jumps up and down.

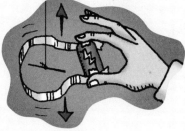

Answers to both magnetic mindmanglers on page 60.

15

BUBBLE, BUBBLE, TOIL AND TROUBLE

Think your chemistry teacher is a bit crackers?
It's always been like that! Chemistry's early days were full of
curious quacks and con men. Here's the science's story...

COMICAL CHEMICALS

Chemical chaos comes in many forms, from disastrous dissolving to spooky sponging, beastly burning to outrageous exploding. Here are some really weird rotten reactions...

LEAF IT ALONE

There are plenty of nasty chemicals knocking around in nature, from bee stings, which are acid, to wasp stings which are alkaline. Bees' acid stings can be treated with alkalines, such as bicarbonate of soda, but don't try this with a wasp sting – it'll hurt more than ever!

One everyday, nasty poison can be found in rhubarb leaves. It's called dicarboxylic (di-car-box-sill-ic) acid. It's there to poison any hungry caterpillar who fancies nibbling it. Luckily, it's not in the stalk, which is the bit we eat with custard as a pudding!

I'VE SUDDENLY LOST MY APPETITE

Soak It and See

Here's something that would come in very handy when you accidentally spill squash on the kitchen floor. A type of sponge called H-spon was invented by chemists in 1974. It's so good at mopping up spills that it can hold up to 1300 times its own weight in moisture. How very absorbing!

I'M IMPRESSED

Bizarre Bombs

• The bang in your Christmas cracker is caused by a chemical called mercuric fulminate. In 1800, its inventor was injured during a lecture showing it off. He must have been crackers (groan)!

• Another explosive is TNT – otherwise known as trinitrotoluene (try-night-tro-toll-you-een). Just one gram of TNT produces about a litre of gases when it explodes! That means it instantly expands to one thousand times its original size! And it only takes a tiny shock to set it off!

• Amazingly, one kilogram of butter stores as much energy between its atoms as the same quantity of TNT! Fortunately, butter tastes nicer on toast and doesn't normally blow up either.

IF THAT'S BUTTER WE'RE ALL TOAST!

HORRIBLY HUNGRY ACID

A super-acid called fluoro-antimonic (flew-or-row-anti-mon-ic) acid has twenty thousand trillion (20,000,000,000,000,000) times the dissolving power of the most powerfully concentrated sulphuric acid. You wouldn't want to dip your fingers in that. They'd be gone in a second!

Sweet Enough Already

Do you take sugar in your tea? Well, if you do, watch what you're stirring in. There's a new type of sugar lump that's 650 times sweeter than ordinary sugar. It's called thaumatin and it's made from seeds of the West African Katemfe plant. If it doesn't give you a lump in your throat it'll certainly help rot your teeth. Now that's what you call a rotten reaction!

KAF! SPLUT!

1. Chemistry is the stinkiest science there is. It often uses killer chemicals like acids to cause a reaction. Some acids are horribly harmful, and the name makes them sound scary, but not all acids are awful; some can be helpful. Aspirin, vitamin C and verruca cream (1a) are all gentle acids.

2. Some photographs – the old-fashioned, non-digital ones – are made using a rotten chemical reaction to 'fix' light-sensitive crystals on paper.

3. Silver-plating puts a layer of silver onto any surface dipped into... an electrical acid bath – so don't be bananas and pick out things you drop in!

4. The dreadfully dangerous 'thermite' reaction is a horribly hot reaction. Rust (iron oxide) reacts furiously with powdered aluminium and turns into pure, molten-hot iron!

5. Talking of horribly hot reactions, welders use a mix of oxygen and acetylene (a-set-uh-leen) gases because they burn together at scorching temperatures.

6. This silly scientist hasn't noticed that he's set his jacket on fire, because he's attached his bunsen burner to the hydrogen tank. Hydrogen burns with an invisible flame.

7. 'Black powder' – gunpowder – produces a little fizz and a lot of smoke when set alight, but light it in a confined space (like in a gun or a firework) and... BOOM!

WHAT'S COOKING?

Anyone can be a chemist. In fact, you may be one without realising it. And if that sounds incredible, consider this: you use chemistry every time you cook. Shocking, isn't it?

How can cookery possibly be chemical? Actually, it would be impossible to cook without chemistry. It's what cooking's all about – from the suspect substances that call themselves school dinners, to the revolting reaction that makes your dad's homemade rice pudding stick to its dish.

Cooking chemicals FACT FILE

THE BASIC FACTS: Most of your food is made up of atoms of a chemical called carbon arranged into larger molecules. Other chemicals are added to improve the taste or texture of the food.

HORRIBLE DETAILS: In the 19th century mysterious things were added to food to save money and make it go further. For example, ground-up bones were mixed into flour. And wooden strawberry pips were added to 'strawberry' jam to make it look more real!

I WISH YOU HADN'T TOLD ME THAT

Kitchen chemistry lab

It's a strange thought, but your kitchen is an awful lot like a chemistry lab.

THE SOUP HASN'T REACHED THE OPTIMUM TEMPERATURE YET

CONCAVE COMBINATION UTENSIL (BOWL)

HEAT ENERGY CONDUCTOR (COOKER)

SPOUTED RECEPTACLE (JUG)

METALLIC CHEMICAL COMPOUNDER (SPOON)

METALLIC SMALL SOLIDS MANIPULATOR (TONGS)

Some machines in your kitchen are mysteriously similar to bits of equipment used by scientists in their labs. Just take the…

Cooker

This is simply a machine for heating food chemicals to produce the chemical reactions that we call cooking.

Then there's the…

WHAT'S THAT SMELL, MUM?

Pressure cooker

This works by allowing water to boil at a higher temperature than usual, so it cooks things faster. It's similar to a lab machine called an autoclave, which is used to kill germs on scientific instruments. And you've probably got one of these…

Thermos flask

Handy for keeping your soup hot or a drink cold on a summer's day. But the flask was originally invented by a clever chemist. In 1892, Sir James Dewar invented the double-walled container to keep certain chemicals carefully cold.

STOPPER →

VACUUM

TWO LAYERS OF SILVERED GLASS

ONION SOUP

So, like it or not – chemistry lessons begin at home!

22

FIVE MIXED-UP FOOD FACTS

1 The burning sensation you get if you eat chilli peppers is due to a chemical called capsaicin (cap-say-kin). According to experts the best remedy for a fiery mouth is a generous helping of icecream! That's tragic!

2 The smell of raspberries found in most yoghurts is due to an added chemical called ionone (eye-on-own). It was originally found in violets. Aah!

3 The bubbles in cooked cake mixture are made by gas! Baking powder contains an acid and a chemical rich in carbon. When they're heated, a chemical reaction produces a gas called carbon dioxide.

4 Vinegar is made from wine that has gone disgustingly sour. This chemical reaction is caused by the waste products produced by bacteria. Yuck!

5 Toast is bread in which the carbon has been partly burned. The smoke that sometimes pours from the toaster when you leave the bread in for too long is made from tiny bits of carbon.

FRESH CARBON – MY FAVOURITE

Bet you never knew!

In 1910, a big shortage of animal fat led to vegetable oils and smelly fish oil being used to make margarine.

SMELLS ALL RIGHT TO ME!

Teacher's tea-break teaser

If you are very brave (or foolhardy) knock on the door of the staff room and try this question on your teacher.

STAFF ROOM

SOME PEOPLE ADD TEA TO MILK AND SOME ADD MILK TO TEA IN THE CUP. IS THERE ANY DIFFERENCE IN THE TASTE AND IF SO WHY?

Answer: It DOES make a difference – and here's why! Milk contains a chemical called casein (case-in). When tea mixes with milk its chemicals break down the casein into smaller molecules. If you add the milk to the tea it means that more casein gets broken down.

This makes the tea taste of boiled milk. That's why chemists 'in the know' add tea to milk and not the other way round!

DEAD DISGUSTING

During the Second World War, German chemists discovered how to make cooking fat from oil – not from cooking oil, but the sort of oil you put in a car! Mmm, tasty!

ONE CAN DOES 2 CARS OR 950 FRIED EGGS

AUTO ÖL

MAKE YOUR OWN SLIME

Slime can't make up its mind – one minute it's oozing like a liquid, the next it's acting like a stodgy solid. Now, let's get going with this goo-tastic gunk…

You will need:
- some cornflour
- a tablespoon
- a large bowl
- 200ml of water in a measuring jug
- a wooden spoon
- a teaspoon
- food colouring

1 Pour some cornflour into the bowl – don't get it everywhere or else the grown-ups get narky!

2 Add a couple of splashes of food colouring to the water. Watch out, this stuff can stain your clothes.

3 Stir some of the coloured water into the cornflour, adding a few drops at a time.

4 Give that wooden spoon some welly and stir the gunk until it starts to thicken up into a paste.

5 Grab a handful of gunk. Watch it ooze out of your hand – it flows like a slippery liquid.

6 Then give the gunk a good squeeze. Wow! Now it wants to be a solid.

DARE YOU DISCOVER... CHEMICAL COOKERY?

Try creating a little bit of chemical chaos in your kitchen with these experimental recipes.

You will need:
- AN ADULT TO HELP YOU!
- 25g butter
- 100g castor sugar
- 75ml water
- a sugar thermometer
- a tablespoon
- a saucepan
- a bowl of ice-cold water
- some chopped apple with skin attached enough cocktail sticks for every bit of apple

YUCKY YEAST

Yeast is no mere chemical. It's ALIVE! Yes – yeast is a tiny fungus like the mould that grows on stale bread. All you do is:

You will need:
- some dried yeast (you can find it in the supermarket)
- two teaspoons
- a tablespoon
- a small bowl
- sugar
- some warm water

1. Mix two teaspoonfuls of dried yeast with two tablespoonfuls of warm water in the bowl.
2. Add a tablespoonful of sugar and stir until it dissolves.
3. Leave the bowl somewhere warm for an hour.

Now it's time to check what's happened to your mixture. You'll find the liquid is frothy and has a funny smell. This is because the yeast eats the sugar and produces alcohol and carbon dioxide – that's the froth.

BLIMEY!

TERRIFIC TOFFEE

Sugar is a complicated compound (combination) of chemicals including carbon, hydrogen and oxygen. Many sweets are simply sugar that's been heated to a particular temperature. For example, fudge is made at 116°C, caramel at 120°C and the hottest of all – toffee. Here's how to make mini-toffee apples. All you do is:

1. Stick a cocktail stick in each of the chopped apple pieces.
2. Mix the sugar, water and butter in the saucepan.
3. Heat to 160°C. Stir it gently. Notice how the sugar turns into a brown, gungy mass.
4. Dip some apple in the mixture. Careful – it's HOT! Then dip the apple into the cold water for 20 seconds.
5. Eat it. And after that there's nothing else for it but to wash up. Never mind, even great scientists have to do this. And luckily, there are lots of chemical cleaners to help you!

PUZZLES

BUNGLING BUNSEN BURNERS

Robert Bunsen was a suffering scientist who kept blowing things up. The Bunsen burner was named after him. Here are six of them! But only two of them are perfectly identical. Which ones? After another accident, Robert can't see too well and needs your help!

A B C

D E F

Answers to both puzzlers on pages 60–61.

GETTING A REACTION!

Here's a collection of substances. All you need to do is work out which reaction belongs to which substance.

a. diamond
b. ozone gas
c. iron pyrite
e. gold

d. soot
f. methyl mercaptan (me-thile-mur-cap-tan)

1. This metal doesn't react with air.

2. Burning wood produces this material.

3. This expensive substance can be changed into cheap pencil lead. In 1905 British King Edward VII was given a large chunk as a present. He said:

> I SHOULD HAVE KICKED IT ASIDE AS A LUMP OF GLASS HAD I SEEN IT UPON THE ROAD

4. German chemist Christian Schönbein (1799–1868) discovered this substance after he noticed an odd smell in his laboratory.

5. This substance produces a sparky reaction when struck by steel. In 1578 explorer Martin Frobisher risked his life to bring this substance from the north of Canada thinking it was gold – it wasn't. What a silly sausage!

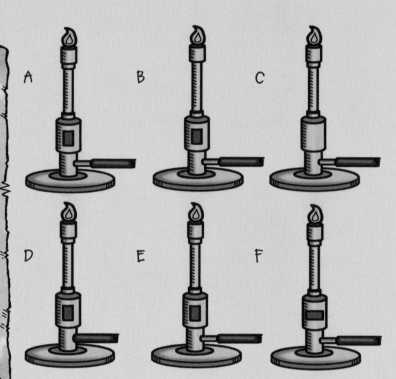

> WE'VE GOT LOADSA GOLD!

> GROAN!

> YEAH! WE'VE GOT GALLEONS OF IT!

6. The human body makes this substance from chemicals in asparagus and it makes the pee really whiffy. In World War Two, US pilots were given asparagus soup to eat in case they were shot down. The pilots were told to pee into the sea and catch fish attracted by the smell.

> MMM, PEE-LICIOUS!

FEATHER FUN

Birds do their silly stuff all over the world and they come in all shapes and sizes. Flying, eating, flirting and nesting are all in a day's work for our feathered friends. We've given extra chores to our long-suffering owl who usually works at night!

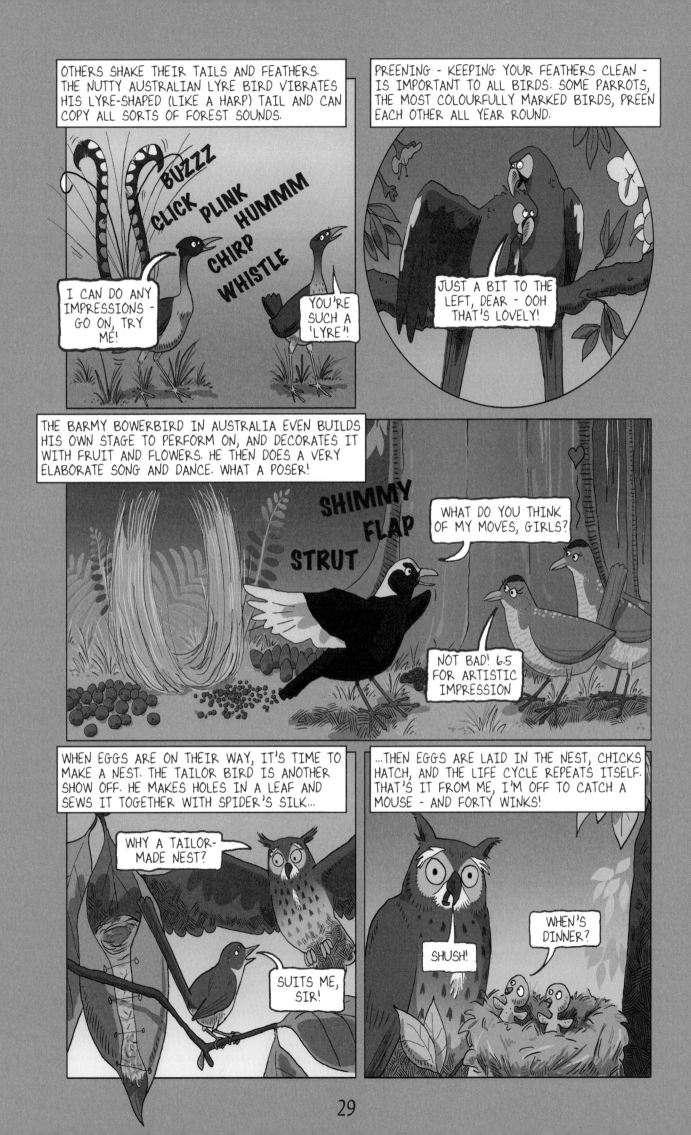

AWESOME EAGLE

With a wingspan of up to 2.5 metres, the bald eagle is an impressive sight when it soars through the skies. But how can such a large creature stay up in the air?

1. The brilliance of this bird begins in its fabulous feathers. They are amazingly light yet strong. The centre bit of each feather, called the shaft, is hollow; the 'vanes' on either side have thousands of hooked branches (1a), called barbs, which trap air between them.

2. Its bones are super light, too. They're also hollow, with needle-like supports inside – like the supporting 'struts' of a bridge.

3. The other special feature that our eagle shares with all birds is that it has fewer bones than most other animals. This isn't because they're missing! They're fused together, so the bird needs fewer muscles (which weigh a lot) to move them. But the eagle's muscles are mega powerful. The strongest ones are its pectoral muscles (3a), which pull the wing downwards and raise the eagle into the air as it flaps for take off.

4. This cross-section of the wing shows it has a shape like a slim slug – pointed at both ends, fatter at the front, with a bigger curve on top. (The shape is called an aerofoil, and it's the same shape as an aeroplane's wing.) As the bird flies, the wing's shape makes the air which flows over the top travel further and faster than the air below, lowering its pressure. The higher air pressure below the wing 'lifts' the bird upwards.

5. Bald eagles spend most of their days gliding on warm-air currents (thermals) that rise up between the mountains and cliffs of their home. From there they hunt for prey, using their keen eyesight to spy fish underwater or rabbits in the grass from over 1km away.

6. The adults return to the nest, or 'eyrie', with live prey, to feed their hungry eaglets. These greedy guzzlers depend on their parents for food for up to six months after hatching.

CRUNCH!

BIRD BONKERS

They're all feathery, flappy, beaky and barmy, lay eggs, make strange noises, and most of them fly. But birds have plenty of differences too. Their anatomy varies in colour, size and shape.

Birds are a colourful and crazy part of life on Earth. Amazingly adaptable, their feathery bodies and wings help most of them rise above many other creatures. But flying isn't as easy as it looks, and learning how can be a tricky process. Guillemots, for example, learn to swim and fly the hard way. Their parents chuck them off a cliff! If they fly – good. If not, they'd better learn to swim!

Meanwhile, mother swallows take food to their chicks but hover just out of reach. If the chicks want to grab their grub then they'd better learn to fly first!

Silly Songsters & Dashing Dancers

Barmy birds make many bizarre sounds, from bell-like warbling to ratchet rifle noises, and do all kinds of dances to attract a mate. Some birds call up to 2000 times a day.

The crane is another rousing entertainer with some great dance routines, going up and under, leaping in the air and trumpeting loud sounds in a dialogue between partners.

Flying ducks

The downward motion of ducks' wings pulls them forward, the most power coming from the outer tips, which move at a higher speed than the section closer to the body.

PLUMED POSERS

While feathers are ideal for flight, very colourful plumage is great for pulling partners. Some of the most colourful feathers belong to birds from South America, Africa and Australia. Here, the cousins of plainer European birds, such as starlings and finches, are crazy for colour. The Gouldian finch of North Australia and the superb starling of East Africa are so colourful it looks as if they've been painted.

I'M PRETTY AS A PICTURE!

The male emperor bird of paradise from New Guinea goes even further with outrageous plumage that looks like candyfloss!

I LIKE TO KEEP THE LADIES SWEET!

I'M AN EYECATCHER FLYCATCHER!

Tails can also be impressive but almost impractically long – this African paradise flycatcher's tail is 20cm and certainly attracts females.

Nifty nibblers

Sometimes birds team up with other animals to get food. When the honeyguide bird spots a wild bees' nest, it makes a special call to attract a passing honey badger. The badger claws open the nest, and can't get stung because of its thick skin. Meanwhile, the bird gobbles up any spare honeycomb. Sweet!

YUMMY HONEY!

The ox pecker in Africa likes to ride on the backs of hippos, zebras and rhinos. The larger animals don't mind. The bird eats the flies that infest their backs. And it even warns of approaching predators. Neat!

KEEP STILL, THERE'S ONE IN YOUR EAR

NO NEED TO SHOUT!

FREAKY BEAKS

Beaks or bills come in a variety of shapes and sizes for all kinds of tasks. Check these out:

The pointy beak of the sword-billed hummingbird is good for poking into flowers for nectar. Some even have tubular or brush-like tongues, too.

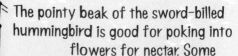
PROD!

hummingbird

CRACK!

toucan

The toucan's powerful beak is very good for crushing tough-shelled fruit and nuts.

DRRRUM!

woodpecker

The woodpecker's strong bill drums into rotten bark, making holes to probe for insects.

GULP!
GULP!

pelican

Pelicans have a 10-litre pouch under their bills – ideal for carrying fish.

Adult beaks aren't just handy for eating or making nests; they're also very useful for chicks. Most birds feed their young by regurgitating their meals down the chicks' throats. (Regurgitation is a posh word for being sick on purpose.) Dutch naturalist Nikolaas Tinbergen (1907-1988) noticed that herring gull chicks peck at a red blotch on their parents' beaks. He set out to discover how important this blotch was.

He made a very crude dummy gull's head, complete with blotch. He also got hold of a dead gull's head and painted out the blotch. Which did the chicks prefer to peck at? The dead gull's head – perhaps it was their dinner? The dummy head with the blotch? Or neither?

the blotch

ERM...

Find out on page 61.

MAKE A SQUAWKY BIRD CALL

Anyone can make this simple bird call. Use it to squawk to the birds!

You will need:
- a piece of paper
- scissors
- your hands

1 Cut a strip of paper 5cm x 0.5cm to make a reed, rather like the one found in a wind instrument such as a clarinet.

2 Stretch the strip very straight between the knuckles of your two thumbs and press them together to hold it in place. The centre of the strip must be free to vibrate in the hollow gap.

gap

3 Blow into the edge of the paper between your thumbs to make a bird call. Once you've mastered it, try different thicknesses and widths of paper to create a range of bird calls.

PEEEEEP!

PUZZLES

Bird-brained Stories

Decide which of these stories is TRUE and which is FALSE!

1. Some birds have really gone batty. The blue-crowned hanging parrot likes to sleep upside down from a branch. Its green back looks like a leaf so it's hidden from hunters.

MORNING!

BLIMEY, A TALKING LEAF!

2. The male bowerbird goes to great lengths to attract a female. First he weaves a plant nest and then paints it with a gooey, blue mixture of blueberries and spit daubed on with a stick, before he fills it with rubbish... irresistible!

3. The African honeyguide bird loves beeswax so much that it flies into bees' nests and risks being stung to death by angry bees.

BUZZZZ

Party Time

Each type of bird has a different silly name for its gatherings. See if you can match the correct bird party with each bird. Here's one to start you off...

A gaggle of	sparrows
A herd of	vultures
A cast of	quails
A brood of	geese
A murmuration of	swans
A host of	chickens
A bevy of	starlings
A venue of	hawks

35

Busy Beaks

Birds' beaks come in all shapes and sizes — long, short, flat, thin (a bit like your teacher's beak, er, nose). Each one is like a tool for catching or scoffing the bird's fave food. Match the beak with a tool and the food it eats.

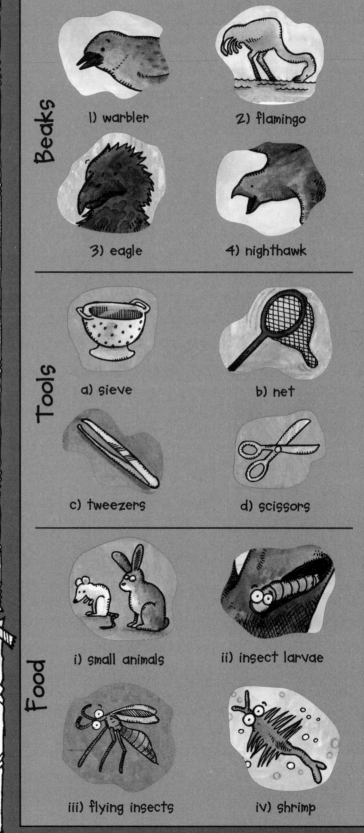

Beaks

1) warbler

2) flamingo

3) eagle

4) nighthawk

Tools

a) sieve

b) net

c) tweezers

d) scissors

Food

i) small animals

ii) insect larvae

iii) flying insects

iv) shrimp

Answers to these brain bafflers on page 61.

BAD BAT CAVE

Of course, birds aren't the only beasts that can fly plenty of insects do it too, but when it comes to being foul flappers, these bats are the stinky stars. Enter this poo-packed home…

1. Guano (say gwar-no) bats (they're named after their poo – called guano, but more of that stuff in a moment!) have very broad ears. As the bats fly around, they make high-pitched squeaks. The squeals bounce of walls and their prey – moths, flying ants and beetles. The bats' big lugholes pick up all the reflections of sound, so they know exactly where everything is in the dark. (It's a super-skill called echo-location.) The bats also have big feet **(1a)** for clinging to crannies. Unlike most other bats, the end of their tails **(1b)** are not webbed – they are 'free', so they're also called free-tailed bats.

2. After a long day sleeping upside down, huddled together for warmth and security, the bats let go, drop into the air and fly out through the cave mouth **(2a)**. Huge, scary swarms of them blacken the late-evening skies **(2b)** as they head off in search of supper.

3. Of course, they need to dump the remains of last-night's digested nosh. Several thousand bats having a quick poo during the day or before they head off in the evening makes for some seriously smelly piles of droppings. The pong comes from ammonium chemicals (also found in human pee). The poo – called guano – also contains carbon, nitrogen and other vital minerals, plus saltpetre – a chemical used to make gunpowder, explosives and fertilizer. Some people actually collect this stuff **(3a)** and then sell it. Ugg!

4. The poo falls into pools and streams that flow through the caves, making a disgusting broth for microbes to grow in. They in turn are food for tiny animals, such as cave crayfish, shrimps **(4a)**, crabs, flatworms, frogs, salamanders **(4b)** and small snakes **(4c)**. Living mostly in near-darkness, these species are often very pale.

5. Some specialised fungi and thousands of species of bacteria thrive on chemicals from the bat droppings too. Cave spiders **(5a)**, whip-spiders **(5b)** and scavenging cockroaches **(5c)** creep about in the dark rocky crevices. Camouflaged moths **(5d)** hibernate in the darker zones.

6. Hunters such as red-tailed hawks keep an eye on the cave mouth and may snatch a passing bat in mid flight. Kestrels and owls also zoom in for the kill. At ground level, raccoons **(6a)** and skunks **(6b)** look for baby bats that have fallen from their perches. They feast on them, and get filthy doing it!

DRIP

5a

5d

MUNCH!

DRIP

PLIP!

PLO

5c

5

SNIF

SNIFF

5b

WILD WOOD

Join the nature-loving Flora Bramble on a terrifying trip to a green and gruesome world. Flora loves a bit of brainy 'botanising', but these peaceful country woodlands hide their true colours… as she soon discovers!

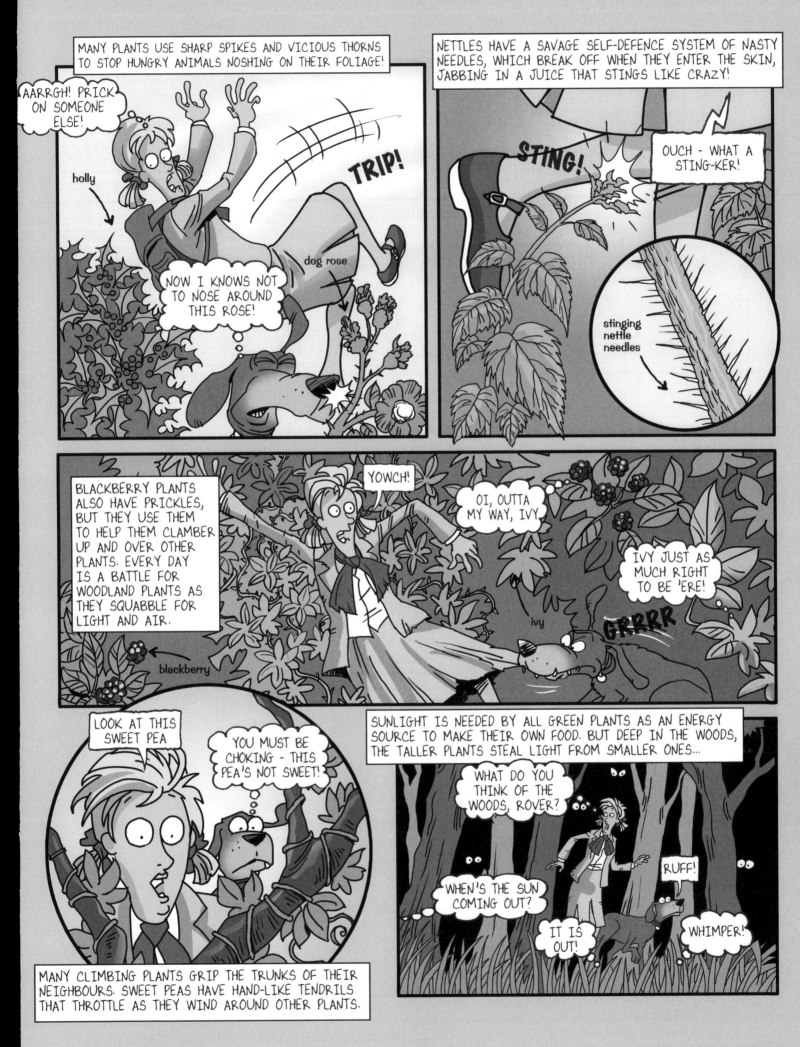

FREAKY THREE

From freaky flowering plants around the world, we've picked
(not literally) the top three craziest and creepiest...

Titan Arum (Sumatra, Indonesia)
Appearance: Just weird. It grows to 3.7 metres high. The sheath that surrounds its huge spike is actually 91cm wide.

Freaky features:
In 1878 Italian botanist Odoardo Beccari found this giant flower growing in the forest. His assistants dug it up, but unfortunately they dropped it! The 1.5-metre-wide corm (swollen root) was full of vegetable fat and exploded everywhere. Eventually Odoardo sent seeds to Italy, and one of the cultivated plants ended up at Kew Gardens in London. When its flowers opened in May 1887 they caused a sensation. Thousands flocked to see them, but the stink from the plant was overwhelming. Several ladies needed first aid because of the revolting reek!

AND HE SAYS I LOOK A FREAK!

Its smell was described as being like rotten fish and burnt sugar. A botanist in the gardens said: "The flowers are pollinated by beetles that eat rotting flesh. They think the smell is heavenly."

Rafflesia (otherwise known as the 'stinking corpse lily') (Borneo, Indonesia). The first European to find it, Sir Thomas Raffles (1781–1826), called it:

THE LARGEST AND MOST MAGNIFICENT FLOWER IN THE WORLD

Appearance:
Giant orange-red cabbage-shaped flower – up to one metre across.

Freaky features: 1. Grows on forest vines and sucks out their juices.
2. Stinks of rotting flesh.
3. It's pollinated either by flies or shrews and then rots into a black stinking mass.

...AND ALSO THE STINKIEST FLOWER IN THE WORLD

Puya plant (Bolivia, South America)
Appearance: Huge head of flowers – the largest in the world. It can measure up to 10 metres high.

Freaky features: Takes 50–100 years to grow to the stage where it flowers. Then it dies. Makes you wonder why the peculiar puya plant bothers.

DON'T LIKE THE COLOUR MUCH

WHY DO I BOTHER?

GREEDY GARDEN

Mrs Twist has let her conservatory garden get out of control. Check out the gruesome growers and super stranglers taking over!

1. Creeping plants love leaking gutters and pipes, and to climb up the side of buildings. This Russian vine can grow up to five metres in just one season!

2. Ivy clings uses roots along its stems to strangle other plants and enter cracked brickwork. It also attracts wasps (2a)!

3. Virginia creeper uses suckers to run rampant.

4. The kudzu vine is very aggressive. With trees to support it, the vine can grow 20–30 metres. It has red or purple flowers which look like sweet peas.

5. This fan palm tree is so strong that it's smashed through the roof and pushed its spiky leaves through.

6. The weeping willow looks harmless, but its roots seek out water, blocking drains and damaging foundations (6a).

7. Hoya climbs using twining stems. Its flowers drip sticky nectar.

8. Citrus trees such as lemons and oranges often get covered with nasty black mould on their leaves.

9. Several sticky carnivorous plants are thriving here. The Venus fly trap likes to snap up snacks all day!

10. Pitfall plants such as the sun pitcher and monkey cup (10a) entice insects to slip inside their flowers.

11. Flypaper plants, such as this sundew, are so sticky that insects can't move once they've landed on them. The butterwort (11a) is good at being a fly's sticky end too!

12. African violets' furry leaves can cause skin rashes. Mrs Twist should watch her feet don't touch 'em!

13. Busy lizzies can get infested with spider mites!

14. Ground ivy is a really rampant grower and is happy to hog the entire floor space of the garden.

15. Miconia also spreads rapidly because it can produce millions of seeds in a short space of time!

16. Meet the stinkers: Stapelia reeks of rotten meat; Russian sage (17) smells like a nasty toilet and Scoliopus bigelovii (18) stinks like a dirty wet dog!

19. Spearmint smells better, but spreads rapidly.

20. The spines of prickly pear can snap off in your flesh – not so good for curious cats!

GROW POTATO TONGUES

You will need:
- an old, wrinkly potato that has sprouted
- a shoe box
- thin card
- scissors
- sticky tape
- a small plant pot
- water
- a little potting compost (optional)
- an adult

Grab a manky old potato and make it grow through a 'maze' so that it looks like gross tongues in a monster face! Plus some celery silliness…

1 First you need to turn your old shoe box into a maze with a monster face. Ask an adult to cut a hole in one end of the box. Draw or paint a gruesome monster face on the end, making the hole the monster's mouth or nose – the uglier the better!

2 Cut two pieces of card so that they are as deep as the shoe box, but not as wide. Tape them into the box as shown. You've made your monster maze.

SPROUT? LOOKS LIKE A POTATO!

HERE'S A NICE SPROUT

3 Grab your potato. It should be such an old spud that it's already started to sprout. Potatoes that have been kept in warm dark places are often the best sprouters. Potatoes sprout from 'eyes' – if your spud has more than one eye, rub the others off.

4 Put your potato in a small plant pot with a little bit of compost and water. Put it in the shoe box, at the end furthest away from the hole and close the lid. Leave the box in a warm place with plenty of sunlight. Now wash your hands thoroughly.

5 Leave your box for about a week (you can take a peep from time to time to see how your spud is doing). Your 'orrible potato shoot will grow towards the sunlight, performing amazing acrobatics to get around any obstacles and break out of the box. Eventually its shoots will reach the sunlight, growing out from the monster's mouth — it'll look as if the monster has disgusting tongue tendrils!

But — here's a puzzler: what colour are the shoots and why? Turn to the Answers on page 61.

IMPORTANT: Do NOT eat the potato or the shoots after the experiment.

TURN CELERY BLUE

1 Add a squirt of blue food colouring to half a glass of cold water. It's best to do this on a kitchen draining board or old table in case you spill any dye — it's very messy and will stain fabrics!

You will need:
- a stick of celery with lots of leafy bits
- blue food colouring
- a glass
- cold water
- kitchen towel to mop up any messy spills

2 Plonk a celery stick or two into the glass. Leave them in a well-lit spot for about a day and watch What happens... ?

3 (a) The celery stick dies...

(b)... or the celery leaves start to turn blue...

(c) ... or the water in the glass turns red?

Answer: (b) The colouring dye is drawn up the stalk and into the leaves, staining them a ghastly shade of blue. This shows how plants take up water from their roots and carry it through narrow tubes (xylem) in their stems to their leaves. The water rises up the stalks by a combination of capillary action — an effect that pulls water up narrow tubes — and evaporation from the leaves that 'sucks' it up. To see how this happens, ask an adult to slice through the stalk with a knife. Look carefully at the cut edge (use a magnifying glass). You'll see little blue spots where you've cut through the water-carrying tubes.

'ORRID ORCHID

In a hot and sweaty jungle in Central America, clinging to a branch high in a tree, there's a fiendish little flower that has a bit of a surprise in store for one species of busy bee...

CRASH!

SLIP

1. The bucket orchid is very pretty — its clusters of small flowers hang from long, drooping stems — but it makes a diabolical trap for one species of nectar–collecting bee, the euglossine (you–gloss–een) bee. The drama starts just before dawn, when the orchid's weird spotty flowers open...

2. Two knobbly glands begin to produce oily nectar, which drips into the steep–sided bucket structure underneath them (2a). After a couple of hours the bucket is full of sweet, delicious and deadly liquid...

3. The trap is now set and the flower wafts a powerful soapy, sweet scent. As if by magic, dozens of male euglossine bees appear, attracted from miles around by the smell. The scent drives them wild! They buzz around the orchid in great excitement, bumping, bumbling and bashing into each other. They swarm all over the mushroom–shaped bulb that grows out of the flower's rim (3a) and begin mopping up the precious oily fluid, stuffing it into 'pockets' on their back legs.

4. Fights start breaking out as the bees squabble over the sweet stuff, and, as they're working on a very slippery surface, accidents are bound to happen (4a)... Sploosh! The bees flounder around in the thick, sticky liquid...

5. The drowning bees thrash around in desperation... but all's not lost. There's an escape route through the tunnel. There's even a little step to help them find it and clamber out (5a).

6. The tunnel is a tight squeeze and the bee has to force its way to the daylight at the exit. But another shock's in store...

7. Suddenly the bee gets clamped by the orchid and held there for several minutes. Meanwhile the sneaky plant is gluing a package to the bee's back. It's a sack of pollen (7a) – the substance used by flowers to fertilize. It has to be carried from one flower to another and the job has been given to this one species of bee.

8. The bee pops out of the exit, shakes himself down and flies off. Eventually he'll fall for the same trick again in another bucket orchid. (They just never learn!) But this time, as the bee squeezes through the tunnel, a special hook in the roof of the exit plucks the pollen sack from his back. The orchid is now fertilized and can go on to develop a fruit and seeds. The bee gets his reward — some scented nectar to attract females.

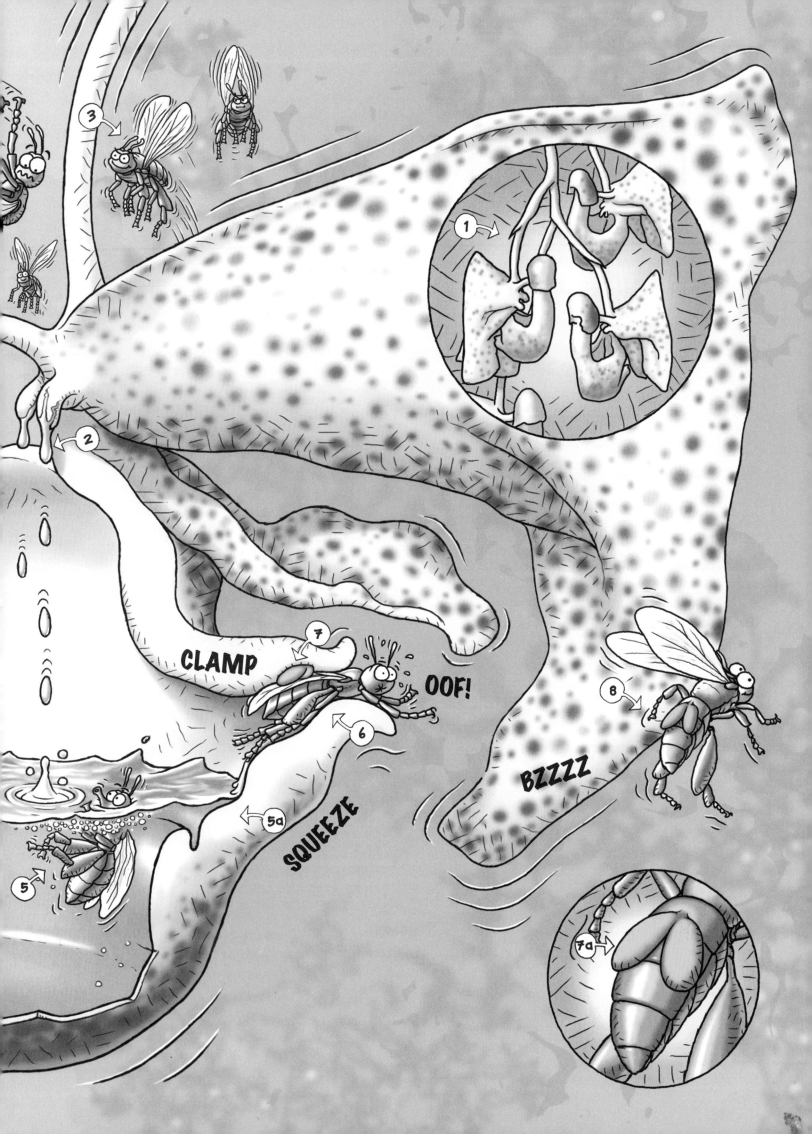

PUZZLES

Vital Vegetable Products

In northern England, in the 19th century, people wove stinging nettle stems into tablecloths! Daft but true!

WHAT A FINE TABLECLOTH...

BEAUTIFULLY WOVEN

ARGH!

So now have a go at matching the following plants with the products that can be made from them...

Plants to choose:
1. Sugar cane
2. Seaweed
3. Wheat
4. Lichen
5. South American bixa tree
6. Cotton
7. Rubber
8. Pine trees

Products to choose:
a). A pair of smelly socks
b). Fuel for cars
c). A nice plate of spaghetti
d). Dye
e). A dirty old wellie boot (Where's the other one gone?)
f). Forecasting the weather
g). Lovely bright orange hair
h). A cardboard box

Peculiar Plant Mysteries

1. You're looking for plants near the Arctic Circle and you find some pretty flowers growing in the decaying skeleton of an animal. Why are they there?

a) The flowers were there first and the animal died on top of them.
b) The flowers were poisonous. The animal ate the flowers and died.
c) The animal died first. The flowers sprouted later and fed on the nourishing rotting bones.

SPOOKY!

2. You find these roots growing on a seashore. Why are they sticking upwards? (Roots normally grow sideways or down.)

a) To breathe air.
b) To stab passing animals and then suck their blood.
c) To stop sand being washed away by the sea and so build up a barrier for the plants to grow behind.

SPIKY!

3. While walking in the European Alps, you find some edelweiss plants. They have very hairy leaves that resemble the fur on a cat's back. What are the hairs for?

a) To attract lots of fleas which pollinate the flowers.
b) To keep the plant warm in the cold mountain air.
c) To make them horrible to eat.

COPY CAT!

4. You cut into the bark of an Australian bloodwood tree and it oozes red juice. Aborigines use the red stuff as a gargle for sore throats. And it sounds like it's ideal for vegetarian vampires!
But what does the tree use the red juice for?

a) To frighten away any creature that bites into the tree.
b) It dries into a gum that protects the cuts from infection as they heal.
c) It sticks the teeth together of any animals that chew the bark.

AND I MEET A LOT OF PEOPLE WITH SORE THROATS

Answers on page 61

STAGGERING SPACE

Earth is a small blue ball floating in a big black sea. And the big black sea is as dark and dangerous as a bad-tempered black bear in a coal cellar! Let's find out how deadly dangerous space really is...

This page is going to hurt your head. There are lots of BIG numbers – numbers so big, they might make your brain blow up. This is because space is a big place. It's also a very scary place, but before we go on, where do we fit into all of this? Ah ha – take a closer look...

Mad Milky Way

You're looking at our very own galaxy – the Milky Way. You can actually see the Milky Way if you happen to live somewhere that's really dark at night – like a place in the country, far away from street lights. It looks like a huge milky drool dribbled across the night sky by a giant baby...

Except it's not milk. It's stars. Something like 200,000 million stars, in fact. And they're bunched together because you're gazing at them from the side (imagine looking at a dish from the side and you'll get the idea...). And this stupendous stretch of stars is 100,000 light years across. (A light year is the distance that light or a radio signal travels in a year – that's about 9.46 million million km, give or take the odd metre or two.)

front view

side view

SLIMY SPACE FACT FILE

NAME: Stuff in Space

THE BASIC FACTS: Here are some of the things that you might see in space...

Galaxies – swirling whirls of billions of stars

ASTRONOMICAL!

Stars – giant balls of super-hot gases. (Our Sun is actually a star.) Even the smallest stars are 80 times bigger than the biggest planets

COOL!

SLIMY SECRETS: Some stars aren't lucky stars if you go too near them. These twinkling terrors could cook you to a puddle of slime.

DARE YOU DISCOVER...
how many stars there are?

You will need:
- a grain of sand
- your finger

Take the grain of sand. Balance it on a fingertip and stretch out your arm. Hold your fingertip against the sky.

Can you still see the grain? Great. Imagine that your eyeballs are space telescopes and your staring at a squitty speck of sky the size of that grain. How many stars would you see?

a) None. The sky is as empty as a piggy bank after a christmas shopping trip.

b) Two

c) 200,000 billion

Answer on page 61

SUPER STAR 'SCOPES

Stargazing has moved on since scientists first gazed at the sky with their naked eyes. Now telescopes can see distant stars using more than just light.

1. Optical telescopes get information by gathering as much light from the sky as possible (**1a**). This optical reflector model is like those really used in Chile and Hawaii. It uses a small convex mirror (bends outwards) (**1b**) and a large concave mirror (bends inwards) (**1c**) to focus the light. The telescope can move sideways or up and down.

2. What's that in the sky? A giant beetle on Mars? No, it's just an insect crawling on some dirt on the mirror (**2a**), and it's seen as a distorted image!

3. This old boffin has a traditional refracting telescope that uses a system of lenses. It's less powerful and also captures less light.

4. A radio telescope like the one at Jodrell Bank, UK, tunes into radio waves (**4a**) from space and captures a whole new set of space images, such as radiation from the Big Bang (**4b**), lobes of a distant galaxy (**4c**), galaxy Fornax A (**4d**), whirling quasar gases (**4e**) and hydrogen in the Milky Way (**4f**). A line or circle of telescopes creates the best image.

5. X-ray telescopes, usually on satellites or space stations, create images of space by picking up very short wavelengths (X-rays) (**5a**). The proportional counter (**5b**) detects the X-rays and converts them to colourful gassy-looking images, such as the remains of a supernova (**5c**), a cluster of galaxies (**5d**) or hot gas pulled into a black hole (**5e**).

6. Space-based infrared telescopes, like NASA's 2003 Spitzer model, detect heat radiation across the spectrum and use it to image entire galaxies (**6a**) or nebulae (clouds of gas in space) (**6b**).

ERM?

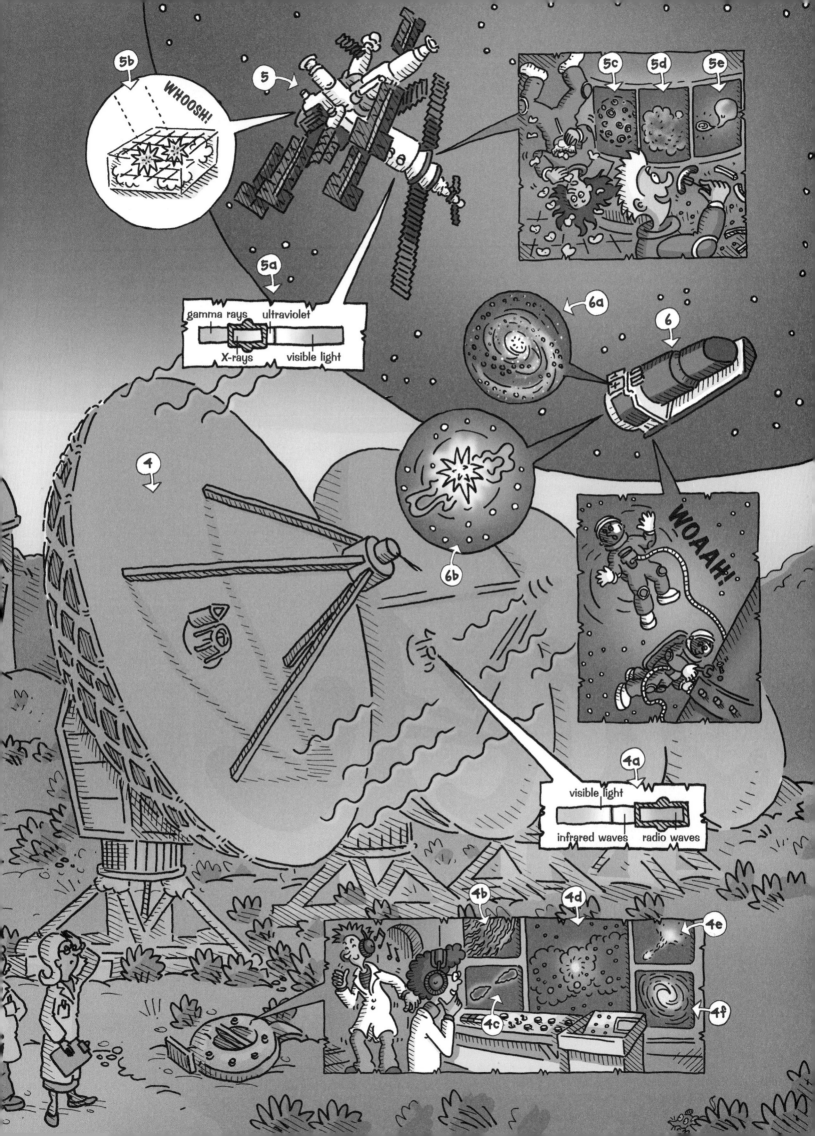

CREEPY CONTACT

Planet Earth drifts slowly around its star, the Sun, stuck in a quiet corner of the galaxy. But its strangely shaped occupants – life forms that call themselves humans – are getting restless and lonely. Will they find new cosmic chums or will it be more horrible than they could ever imagine?

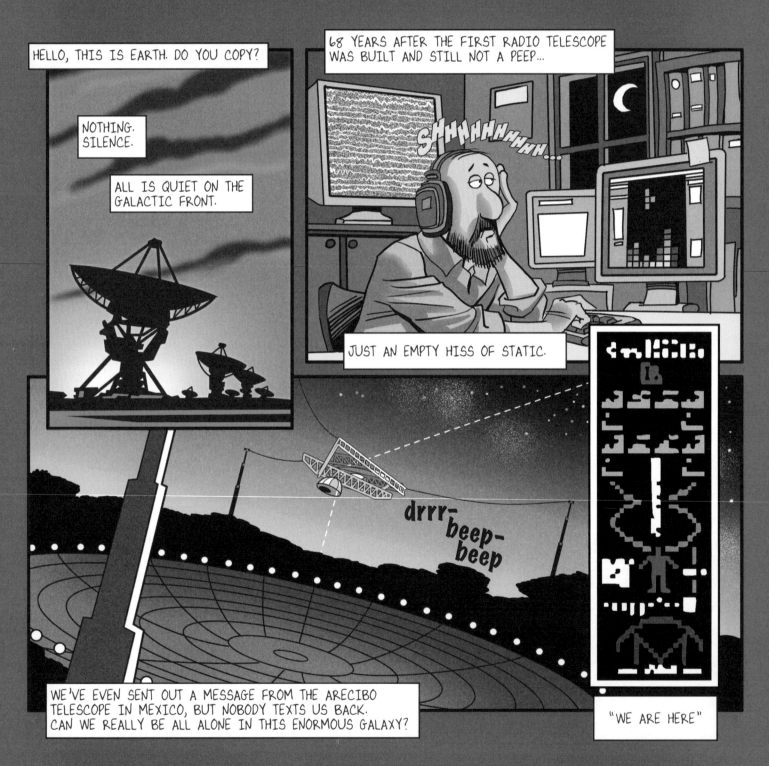

HELLO, THIS IS EARTH. DO YOU COPY?

NOTHING. SILENCE.

ALL IS QUIET ON THE GALACTIC FRONT.

68 YEARS AFTER THE FIRST RADIO TELESCOPE WAS BUILT AND STILL NOT A PEEP...

SHHHHHHHHHH...

JUST AN EMPTY HISS OF STATIC.

drrr-beep-beep

WE'VE EVEN SENT OUT A MESSAGE FROM THE ARECIBO TELESCOPE IN MEXICO, BUT NOBODY TEXTS US BACK. CAN WE REALLY BE ALL ALONE IN THIS ENORMOUS GALAXY?

"WE ARE HERE"

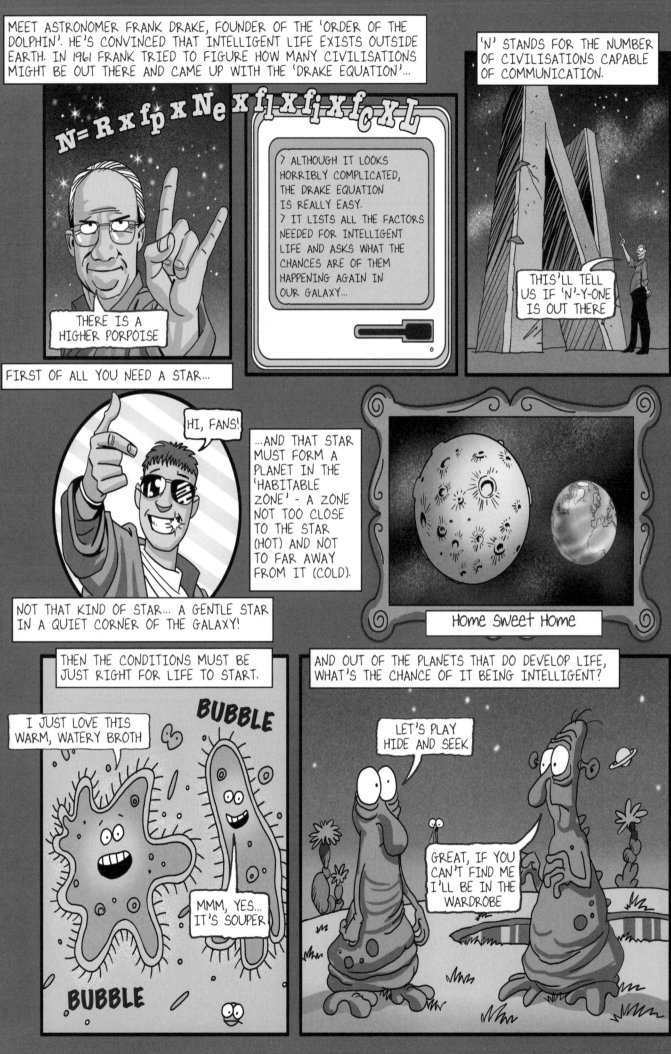

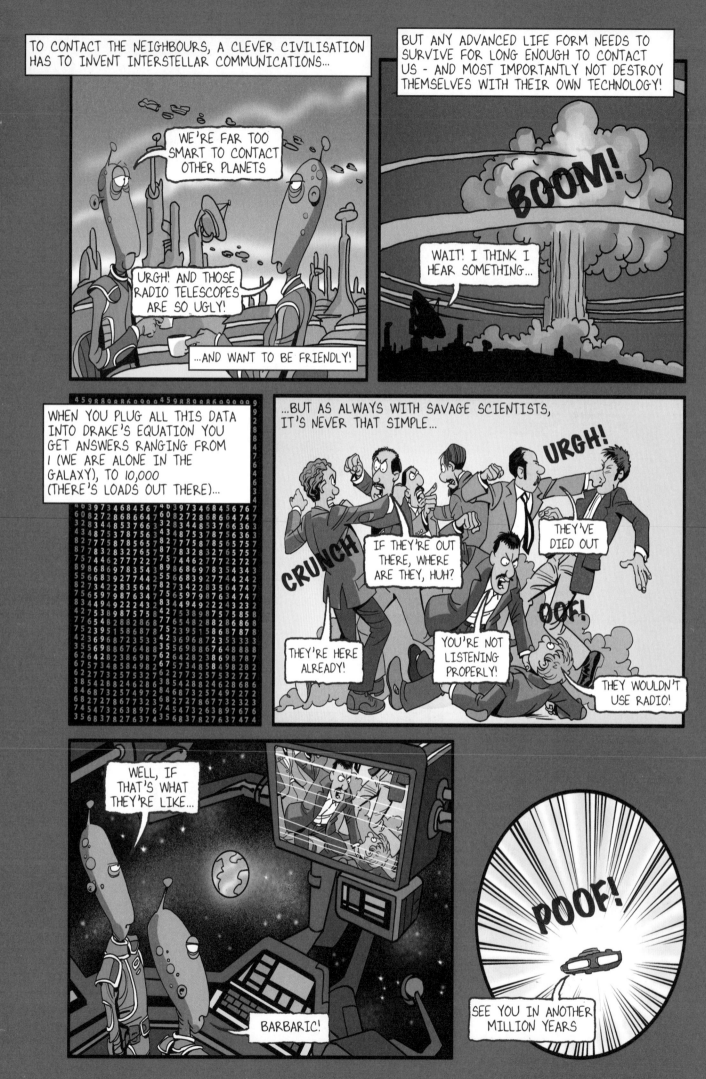

PUZZLES

Could You Face Space?

Slobslime, a dotty Snotty from the planet Splott, has foolishly gone on a space walk without her spacesuit. Which of the following things would she REALLY experience?

1) Her skin would freeze solid
2) Her skin would burn to a crisp
3) Her hair would turn purple
4) Her eyeballs would go pop
5) Her lungs would burst
6) She would begin to sing the Splott national anthem
7) Her blood and body juices would begin to boil like slimy green soup
8) She would feel sick

STARTLING SPACE QUIZ

1. If you stayed in bed all day at a point on the equator (the imaginary line around the Earth's middle), what would happen to you?
   a) In one day you'd travel a distance equal to going around the world
   b) You'd get dizzy
   c) Gravity would pull you up into space

2. If you stayed in bed at the North Pole, what would happen?
   a) You'd be sucked into the Earth's core
   b) Your bed and you would go round in a little circle like an ice skater in a spin
   c) Your bed would take off

3. Which of these facts about the Sun is true?
   a) It sings
   b) It turns off at night
   c) It gives off radio waves

Are We a Clone?

Two of the spaceships below are identical. If you don't get in a spin, you'll be able to spot them.

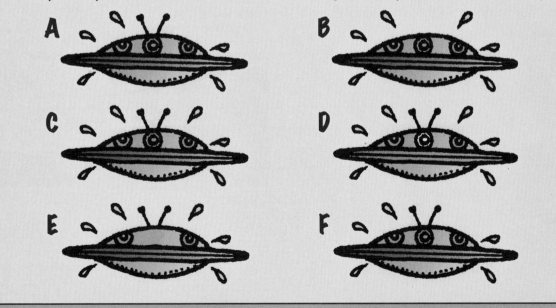

Answers on page 61

MAKE A BRILLIANT ASTRO BOX

Imagine stars are dots in lots of dot-to-dot puzzles. The shapes they make are called constellations, and they're a handy helper for astronomers. Here's how to make your own constellation show!

You will need:
- a shoebox
- four pieces of black paper
- a pencil
- a ruler
- scissors
- a bright pen
- Blu-Tack
- sticky tape
- a torch (preferably two)

1 Take the lid off your shoebox (remove any shoes inside) and draw a rectangle on the inside of the lid – leaving a border of about 2cm all round. Cut out the rectangle.

TOP TIPS

If you don't feel confident about copying the constellations, enlarge them on a photocopier by 300% and trace them onto each sheet.

2 Cut the sheets of black paper to fit inside the shoebox lid. Now copy a different constellation (see right) onto each of your pieces of black paper. Drawing a grid onto the black paper will help you position the stars correctly. (A grid with bigger squares allows you to enlarge the shape to fill the space.)

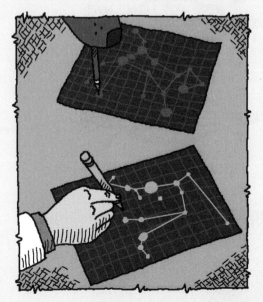

3 Using a sharp pencil, make a hole where you've drawn each star. Place a blob of Blu-Tack underneath each star as you do it, to avoid spoiling your parents' kitchen table. Heaven's above – you've made a constellation!

4 Put four little strips of sticky tape on each side of your constellation and stick it inside your shoebox lid. Now turn on the torch (or two torches if you have them) and place it inside the box.

5 Now it's time to trip the light fantastic — close the curtains, turn off the light and enjoy! When you've seen enough of these constellations, visit your local library and find a good book on astronomy to find more.

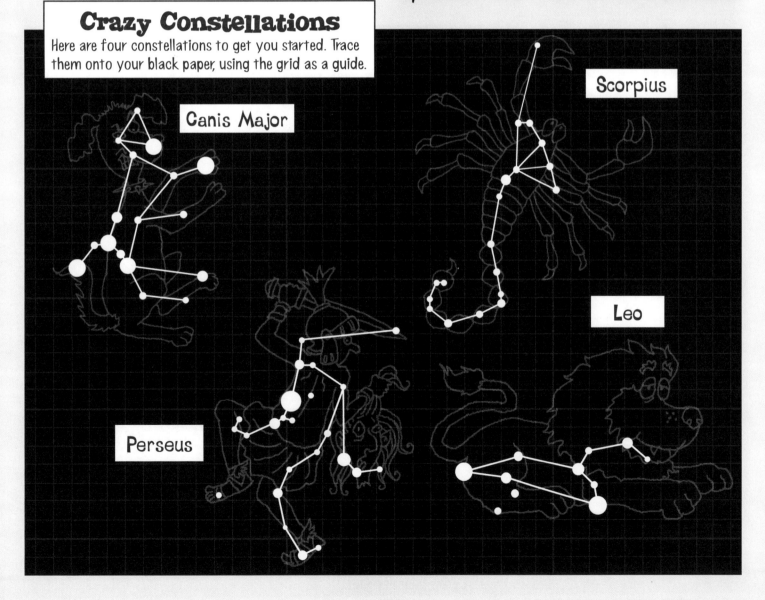

Crazy Constellations

Here are four constellations to get you started. Trace them onto your black paper, using the grid as a guide.

Canis Major

Scorpius

Leo

Perseus

WEIRD WORMHOLE

Ever wondered how you might travel to another universe in super-fast time? Some scientists think this is theoretically possible - through a 'wormhole'!

1. Space is very big — so big that even light can take a long time to travel across it. Years in fact! However, a light year is not an amount of time, but the distance light will travel in a year. Light travels very fast, at 299,792,458 metres in just a second, so in a year it travels an amazing 9,460,730,472,580.8 kilometres (that's well over nine million million)!

2. Even with our intrepid astronaut's super-fast spaceship, a journey to another galaxy will take him an extraordinarily long time. He gets very tired and bored **(2a)**. On his own clock, lots of time will pass, and he is starting to age **(2b)**. Will he live long enough to get to his destination?

3. There is, in theory, an alternative — to pass through a wormhole! No one has ever seen one but they COULD look something like this. Imagine a worm eating its way to the other side of an apple. To take a shortcut, it must burrow right through it rather than go around the edge. So in space, first the astronaut must enter through the wormhole's mouth **(3a)**.

4. The spaceship will now travel through a hole that's kept open by negative energy, also known as 'exotic matter'. But wormholes are known to be unstable, and may collapse at any time. Another unfortunate spaceship has been crushed when this happens. Yikes!

5. Fortunately, our now rather terrified traveller is whooshing his way to the other side of the wormhole, having survived through the middle area known as the throat. Luckily it didn't swallow him up!

6. Some wormholes lead to another part of the same universe, but this one - of a type called an inter-universe Lorentzian traversable wormhole (BLIMEY!!!) causes the astronaut to emerge in another universe entirely. The place is inhabited by scary space aliens, but they are just as terrified as him **(6a)**!

7. Our traveller has passed through a huge distance in almost no time at all. He hasn't aged a bit!

8. But if it were possible to meet his other self who has taken the long route, he might find he's a very old man!

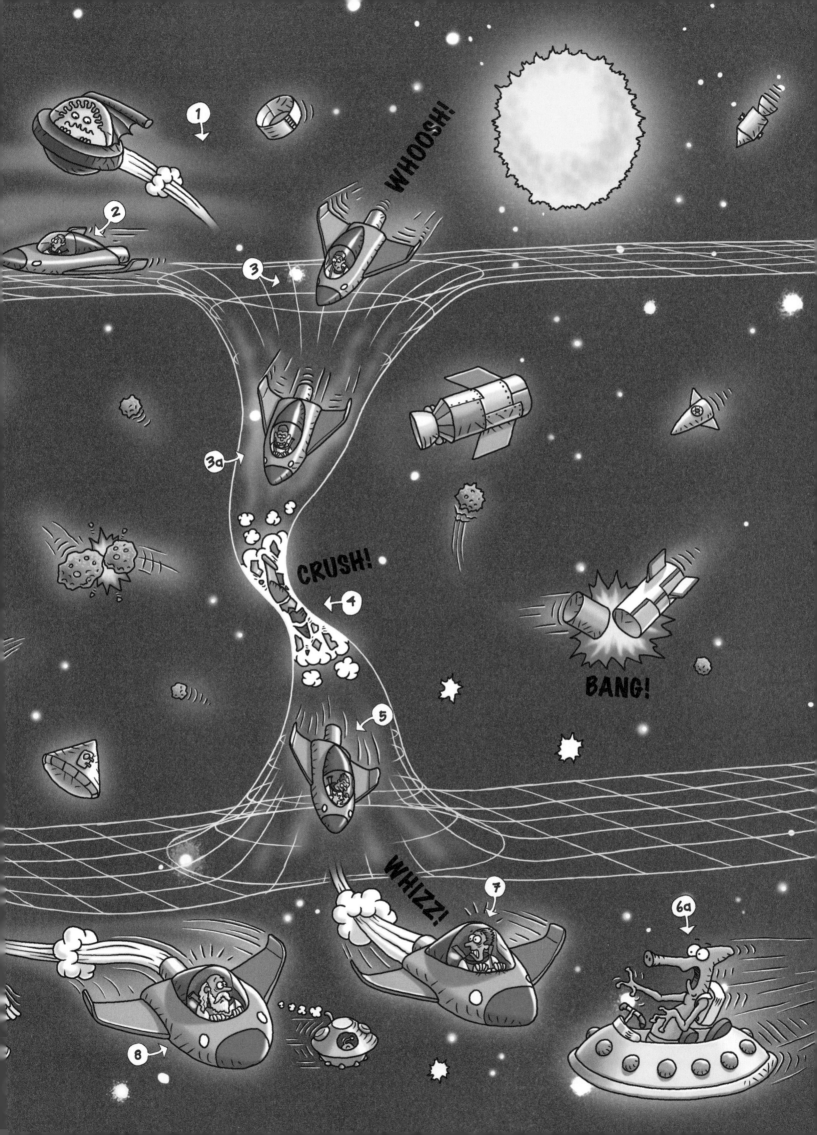

PUZZLE ANSWERS

So are you a supercharged magnet mind or a bird-brained vile vegetable?
Find out here... and learn why those potato roots are so revoltingly pale!

DARE YOU DISCOVER... HOW TO MAKE A MAGNETIC PLANE? P6

b) The closer you are to the magnet the stronger its force. The area around a magnet that is affected by its force is called a 'magnetic field'. (Mind the magnetic cow pats!)

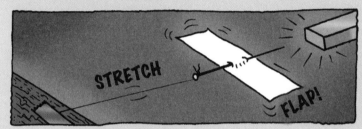

DARE YOU DISCOVER... IF MAGNETISM WORKS UNDERWATER? P7

a) This proves magnetism works through glass AND water.

TRUE OR FALSE? P14

1. True. 2. True. 3. False.

THE CASE OF THE MURDERED MAGNET P15

b) Bashing it with a hammer. The clue – the metal was ice-cold. Both the flame and the electric current would have worked, but they would have warmed the magnet! The bogus clue was d), of course. But what a way to go!

MAGIC NEEDLE P15

b) The needle twists around. The magnetic field produced by the loop of foil keeps pushing away and then attracting the magnetized needle, just like a real electric motor.

BUNGLING BUNSEN BURNERS P26

The two identical burners are a) and e).

CHOCO DYNAMO P12-13

Here are all those hidden choc bars devised by dastardly Dr Chocolato and his child-powered dynamo!

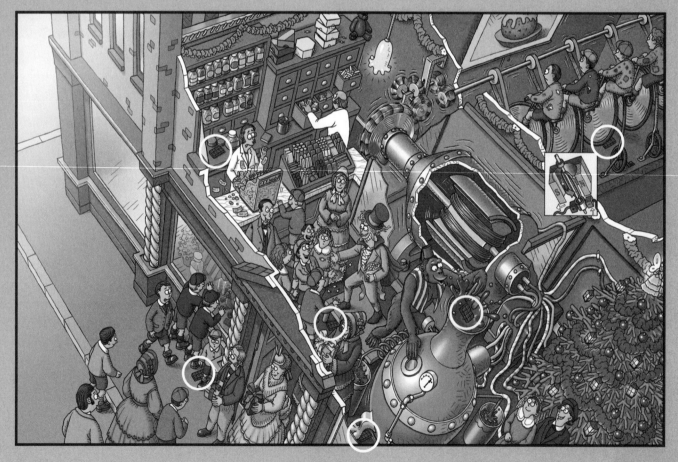

GETTING A REACTION P26

1 e) Just a few metals don't react with the oxygen in the air, and they include gold, platinum and silver. Iron reacts with oxygen to form rust.
2. d) Soot is unburnt carbon produced when natural materials, such as wood, burn.
3. a) In 1905 Edward was given the world's largest diamond as a birthday present. (What an ungrateful grump!) All diamonds eventually return to a more stable form of carbon – graphite (the same stuff found in pencils) – but this reaction is *extremely* slow.
4. b) Ozone means 'I smell' in Greek!
5. c) Iron pyrite is known as 'fool's gold' because its brassy yellow colour fooled many people into thinking it was gold.
6. f) Methyl mercaptan. It reeks of rotten cabbage, garlic, onions, burnt toast and stinking blocked toilets!

FREAKY BEAKS P33

Would the sight of a human head make you feel hungry? No! The chicks pecked at the dummy head instead. Tinbergen went on to prove that the red wasn't important; any colour was OK as long as the patch was clearly visible.

BIRD-BRAINED STORIES P35

1. TRUE.
2. TRUE. The bower bird will decorate its nest with dead insects, bottle tops and bird skulls.
3. FALSE. It gets a honey badger's help, remember?!

PARTY TIME AND BUSY BEAKS

1 + c + ii; 2 + a + iv; 3 + d + i; 4 + b + iii

PARTY TIME

A gaggle of geese. A herd of swans. A cast of hawks. A brood of chickens. A murmuration of starlings. A host of sparrows. A bevy of quails. A venue of vultures.

GROW POTATO TONGUES P45

The stems of plants that thrive in sunny places grow towards the light. This is controlled by a hormone that makes the cells on the stem's shady side grow faster than on the sunny side, so the stems bend upwards. The same hormone helps them grow over obstacles. In sunlight, the green pigment chlorophyll develops in the leaves and stems that helps the plant to make food. When the plant is in darkness, the green pigment doesn't develop so the shoots are pale yellow.

VITAL VEGETABLE PRODUCTS P48

1. b) In Brazil, sugar cane is made into alcohol and this is used as a fuel for some cars.
2. f) Laminaria seaweed feels dry and brittle in dry weather but as rain draws near and the air grows moist the seaweed takes in water and feels sticky.
3. c) Spaghetti is made out of pasta which is made from semolina… which is made from ground-up wheat.
4. d) Lichens were used to make traditional dyes. A typical recipe involved leaving the dye to rot in a mix of stale pee and a chemical called slaked lime.

5. g) Native South Americans use bixa seeds to make groovy orange hair colouring. The juice also keeps mosquitoes away.
6. a) The cotton in your socks is made from hairs that help disperse the seeds of the cotton plant. The hairs form inside the seed pods and are spun using machines to make the cloth.
7. e) Wellington boots were traditionally made from rubber – the congealed juice, or latex, that oozes from the rubber tree when you cut its bark.
8. h) Wood pulp from trees such as pines is used to make cardboard.

PECULIAR PLANT MYSTERIES P48

1. c) The soil is very poor in this region but the rain has washed nourishing minerals from the dead body into the soil. So the flowers flourish. The skeleton also protects the plants from the biting wind. There could also have been some plant seeds in the animal's guts.
2. a) They're a type of mangrove root. They stick out of the water to take in air. When the tide comes in they stop taking in air and the plant has to hold its breath until the tide goes out.

3. b) The hair protects the plants from freezing, just like a fur coat.
4. b) It dries into a sticky gum to seal wounds.

DARE YOU DISCOVER…

c) There really are 200,000 billion stars in every grain-of-sand-sized bit of sky!

COULD YOU FACE SPACE? P55

Either 1 or 2 would happen. Away from a star, like our Sun, space can be colder than -100° C, but in sunlight it can be hotter than 120° C; 3 is not known to happen, although Snotties are strange creatures (it wouldn't happen to us anyway!); lack of air pressure would make 4 & 5 happen; Slobslime may well do 6, but it's not scientifically necessary; 7 would happen due to space radiation; and 8 is bound to happen, although it would come out in floating globules and look very nasty.

ARE WE A CLONE? P55

A & F are the same

STARTLING SPACE QUIZ P55

1. a); 2. b); 3. a) & c) Yes, it REALLY sings – as its surface wobbles in and out, waves of gases like sound waves, ripple across it. But they won't reach us here, unless the Sun starts a radio station!